PLANNING FOR DISASTER

A Guide for School Administrators

SECOND EDITION

HARMON A. BALDWIN

Published by
Phi Delta Kappa Educational Foundation
Bloomington, Indiana U.S.A.

Cover design by
Victoria Voelker

Phi Delta Kappa Educational Foundation
408 North Union Street
Post Office Box 789
Bloomington, Indiana 47402-0789
U.S.A.

Printed in the United States of America

Library of Congress Catalog Card Number 99-68091
ISBN 0-87367-818-4
Copyright © 1999 by the Phi Delta Kappa Educational Foundation
Bloomington, Indiana

All rights reserved

Table of Contents

Preface to the Second Edition 1

Preface to the First Edition 3

Disaster and Emergency Preparedness 5

Emergency Preparedness Plans: A National Overview 7

Time to Get Started 11

A Potpourri of Ideas from Selected Emergency/
Disaster Plans 15

Five Steps to a Safe and Secure School 19

Resource Addresses 21

Appendix 23

Preface to the Second Edition

The decade since the first edition of this guide was published has witnessed an inordinate number of school tragedies. These have ranged from such natural disasters as tornadoes and hurricanes to human horrors, such as the terrorist-style attack at Columbine High School in Colorado and a rash of other school shootings throughout the nation.

Disasters occur without warning. After all, if we had ample warning, there would not be a disaster. Thus schools must be prepared for the unexpected. When the tornado is at the door or the shots are being fired, it is too late to plan; it is time to act. A well-made disaster preparedness plan is essential for educators to act well.

Given the events of the last decade, schools currently are striving to develop new disaster preparedness plans. These schools should learn from the past, from the experiences of schools that already have solved many of the problems that arise when developing such plans. The information in this guide was compiled by Harmon Baldwin, a long-serving district superintendent, from 38 emergency/disaster plans used in schools across the nation.

This second edition of *Planning for Disaster* has few changes from the original. The planning suggestions compiled a decade ago are still current and still sound. Our hope is that school administrators will use this guide to prepare themselves and their staffs, students, and communities for that moment one hopes will never come — when disaster strikes a school.

> Donovan R. Walling
> Editor of Special Publications

Preface to the First Edition

This booklet is intended to provide school leaders with information that can be used in developing an emergency or disaster plan suitable to their particular school or school district. Phi Delta Kappa's interest in this topic was triggered by a number of recent incidents in which school people were faced with crisis situations that endangered the lives of both students and school personnel.

In pursuit of this interest, Phi Delta Kappa asked its members to provide copies of school emergency/disaster plans currently in use in their communities. Thirty-eight plans were received and subsequently analyzed. From this analysis, it became clear that no single plan can serve all schools or school districts. Any plan must be tailored for a particular school, community, or region of the country.

In this booklet the reader will find information useful in developing a specific emergency/disaster plan for a school or district. It provides an overview of the problem, identifies steps to be followed in developing a plan, and suggests what the content of the plan should include. The Appendix includes a sample emergency/disaster plan that can be used as a starting point.

While the information provided here should be helpful, those responsible for developing an emergency/disaster plan should realize that it is a time-consuming and ongoing activity. Those responsible for developing and implementing the plan must be fully aware of their responsibilities and of the time required to complete the plan. The time will be well spent. Having an adequate plan in place cannot prevent an emergency or disaster from happening. It will, however, help to ensure that the turmoil and suffering are greatly minimized.

Harmon A. Baldwin

Harmon A. Baldwin is a retired superintendent of the Monroe County Community School Corporation in Bloomington, Indiana.

Disaster and Emergency Preparedness

You are the school principal. It is 7:45 a.m. Students and faculty are gathering to start another school day. As you stand in the entry foyer warmly greeting them, a student, obviously distressed, dashes down the hall and reports that a teacher is involved in a fist fight with a student, and the student has pulled a knife. Do you have established procedures to handle such an incident?

You are the school principal. Morning recess has just ended. A student bursts into your office shouting that some woman just shot a boy in the restroom and was last seen entering a classroom with a pistol drawn. Do you have established procedures to handle such an emergency?

It is recess time and the children on the playground are busily involved in kick ball, sliding, swinging, running, and other favorite recess activities. An appropriate number of teachers or aides are supervising the activities at various locations around the playground. Suddenly and without warning, your playground is strafed with rounds fired from what is later identified as an AK-47 assault rifle. Where minutes earlier happy children played, five now lie dead and many others, including one teacher, lie wounded. Do you have procedures for handling such a situation should it happen in your school?

These three incidents actually have occurred in U.S. public schools recently, and other schools have experienced similar crisis situations. For years parents have sent their children to school without undue concern, believing that the school was a safe haven. Have crisis incidents such as those described above denied schools of their image as safe and secure environments? Is there no sanctuary from random violence? Can school administrators develop and implement security plans that provide adequate protection to children and faculty?

This is the challenge facing the leaders of America's schools. No longer can boards of education, superintendents, or principals ignore the possibility that their school could be next. School officials must prepare for the unthinkable. Violence and tragedy have struck before in unlikely settings — it could happen again!

It is imperative that local school officials assess the adequacy of their existing emergency/disaster plan (if they have one). This aspect of school administration cannot be accomplished by complacently filing a copy of the plan in the superintendent's office. At least annually, there should be an administrative review, faculty discussion, and appropriate updating to cover new situations.

Emergency Preparedness Plans: A National Overview

Many school districts are currently developing local disaster and emergency plans in response to existing state regulations or laws, not as a result of actual on-site emergencies. For example, the Indiana State Board of Education, as part of its school accreditation standards, requires that each school or attendance center provide emergency-preparedness instructions to students at all grade levels annually. The regulation sets forth a list of possible emergencies including fires, tornadoes, nuclear disasters, winter storms, flash floods, and earthquakes. However, the current regulation does not require planning for such traumatic events as befell the Hubbard Woods Elementary School in Winnetka, Illinois, where a student was shot, or the Cleveland Elementary School in Stockton, California, where the playground was strafed. Many other states have regulations similar to Indiana's. However, complying with your state's regulation does not preclude the alert administrator from going further and preparing a disaster plan tailored to meet any eventuality.

Some states have begun to develop separate initiatives to respond to the growing threat of violent disasters in schools. Their efforts range from changes in planning to updating educational codes. For example, the state superintendent of South Carolina has appointed a Safe School committee, which has been charged with recommending programs or practices to improve the safety of schools in that state.

On 7 April 1989, the New York State Board of Regents enacted a rule that requires each board of education in that state (other than New York City) and each board of cooperative educational services (BOCES) to develop a school emergency management plan by 1 October 1990. The Board of Regents' rule defines a series of emergencies to be included in each local plan. In addition, the rule stipulates that each local plan be updated annually.

The federal government also endorses the idea that schools should provide a safe environment. In 1981 the Attorney General's Task Force on Violent Crime recommended that the attorney general exercise leadership in building a national consensus that crime, violence, and drug abuse have no place in a school. He was urged to support vigorous local law enforcement efforts when school conditions warranted. In 1982 the President's Task Force on Victims of Crime recommended, among other things, that school-based crimes be promptly reported and that support be provided to student victims. In 1983 the National Commission on Excellence in Education stated in its influential report, *A Nation at Risk,* that there were both disruptive students and disruptive circumstances affecting public schools, and offered recommendations for dealing with them.

As a result of the recommendations of these federal panels, the Department of Justice and the Department of Education developed a demonstration program to deal with problems of school safety. Originally called Safe Schools-Better Students, it was field-tested in Anaheim, California; Jacksonville, Florida; and Rockford, Illinois. The program since has been refined and renamed SMART (School Management and Resource Team).

SMART has five major components. They are:

1. Commitment: The school superintendent and each school principal must be committed to the goals and methods of the program.
2. Safety and Security Audit: This audit includes an in-depth analysis of school district policies and practices with regard

to student/faculty safety, discipline, crime, and drugs.
3. Incident Projecting System: This system clearly differentiates between disciplinary infractions and criminal acts in schools; it uses school district computers to analyze data describing patterns of disruption and crime.
4. SMART Teams: Teams comprised of representatives of all groups involved in the local project meet and review incident reports each month, set priorities, develop intervention plans, and monitor results.
5. Inter-Agency Coordination: Such coordination involves school and community leaders who deal with principals either directly or indirectly and who concentrate their efforts on youths who commit crimes on school grounds.

With the national and state efforts to help local schools to provide a safe environment, no school leader needs to develop or update an emergency preparedness plan in isolation. However, school leaders should remember that each school or district must have a plan tailored to its specific needs.

But are school leaders in the United States really committed to providing safe and secure environments in schools? If Indiana school administrators are typical of their cohorts in other states, the answer may be no. In a recent communication to Indiana school superintendents, Donald D. Donath, chief inspector of the Indiana Department of Fire and Building Services, said, in part: "This office ran a computer readout of the ten most cited violations [of Indiana's law] in the past year. To my dismay, number nine was not having correct records on file. . . . Further investigations ... show this problem is *being ignored* in consistent violations over the last three years. . . . We understand the inconvenience of ... drills, but we feel the possibility of [student] death or injury outweighs [any inconvenience]" (emphasis added). The possibility of death or injury should be sufficient motivation to have a well-developed emergency-preparedness plan that the staff and student body have been trained to implement.

Time to Get Started

School security consultants state that the best defense against random violence is a comprehensive school security plan that is as concerned with routine fist fights among students as it is with the attack by an armed intruder. So how does a school leader start to develop a plan that is tailored to the unique needs of a specific school in a specific community? Although not intended as a blueprint for all schools, the suggested steps that follow will start you on the way to an emergency-preparedness plan for your school or district.

1. *Gain commitment from the board of education.* The commitment of the school board is best demonstrated by the adoption of a general policy that deals with potential emergencies. Having such a policy in place serves both to direct and support the administration as it develops and/or updates plans for handling potential crises. A sample general policy statement with accompanying regulations is included in the Appendix.

2. *Conduct a security audit.* Most states have laws or regulations that direct local schools to conduct drills to reduce the likelihood of injury or loss of life from fires, tornadoes, or earthquakes. Too often, however, state directives narrow the focus of local school officials to the more common emergency situations. A security audit allows the alert administrator to look beyond the most common emergencies to the unexpected risks and potential hazards.

Although a security audit can be done by the staff members, they may be too close to the situation and overlook potential hazards. One option is to invite local law enforcement officials to conduct a security audit. In addition to their general expertise in matters of public safety, conducting a security audit gives them the opportunity to become acquainted with the special features of the schools in their communities. Another option is to secure outside consultants to conduct the audit. The National Alliance for Safe Schools, for example, has experience in conducting school security audits. (See Resource Addresses.)

3. *Develop your plan.* The best emergency-preparedness plan will require the cooperation and concern of everyone in the school or district. Therefore, representatives of all school groups should be involved in the development of the local plan. Parents and other community members interested in safe and secure schools should be invited to serve on the committee that develops the plan. Impress on the planning committee that school security is *everyone's* business. Involvement in developing the plan leads to a sense of ownership. Ownership develops responsibility. Responsibility leads to a safer environment.

One of the first issues a planning committee may take up is controlling access to schools. Our schools long have prided themselves on being open and accessible places. But in today's world, can schools be open and accessible and still be secure? The ideas of a planning committee on controlling access to a school may range from indifference to paranoia. Often a school security expert can advise you on a middle ground to take that provides a degree of control but avoids a school-as-fortress mentality.

It may be necessary to remind the planning committee that most emergency situations in a school result from day-to-day incidents within the school, not from the actions of outside intruders. A good school emergency plan will incorporate procedures that prepare students, staff, and others to respond rationally regardless of whether the situation calls for a routine fire or tornado drill or a true emergency requiring the evacuation of the building or the sealing off of an area.

Below are the steps the committee might follow in developing an emergency-preparedness plan for a particular school or district.

- Draft a policy statement on disaster and emergency preparedness and present it to the school board for its review and subsequent approval.
- Identify those who are to be involved in preparing the plan and give them their assignments.
- Conduct a security audit to determine the unique conditions or circumstances of the building, site, and neighborhood that must be addressed in developing the plan.
- Collect and review existing emergency/disaster plans in other school districts and assess their applicability for this community.
- Conduct a survey to determine the community resources, both human and material, available for developing and implementing a plan.
- Analyze existing school policies in the area of student/faculty safety, discipline, crime, and drug use to determine their relevance to the emergency/disaster plan.
- Review state laws or regulations to determine their applicability to and impact on the school's emergency/disaster plan.
- Prepare a tentative plan tailored to the unique circumstances of the school or district, circulate it, and request reactions and input.
- Submit the plan to one or more experts not directly involved in its development and ask them to review it and comment on its adequacy.
- When the tentative plan has been drafted, review to make sure that it includes:
    a. procedures for the evacuation of students,
    b. procedures for communicating with parents and the community,
    c. procedures for dealing with news media,
    d. descriptions of responsibilities for all personnel,

  e. procedures for training those involved in implementing the plan, and
  f. provision for regular updating.
- Secure approval for the final plan and begin implementation procedures.
- When you read or hear about a school disaster in some part of the country, gather all the information about it from the media and other sources and use it as a case study or simulation exercise to test the adequacy of your emergency/disaster plan.

4. *Keep your plan up to date.* Once your plan is in place, do not expect it to remain unchanged. School security is in constant flux. To maintain safe schools, administrators must review their emergency/disaster plan on a systematic basis to ensure that it reflects changing conditions. Also, training in implementation procedures will need to be repeated as principals and faculty turn over and as the student population changes. Another way of keeping up to date is reading the literature on school safety and school security available from various agencies.

A Potpourri of Ideas from Selected Emergency/Disaster Plans

Many good ideas were gleaned from analysis of the 38 emergency/disaster plans that schools submitted to Phi Delta Kappa. Some representative ones are included here for school leaders to consider as they develop their own plans.

- Many plans were published as a manual; some used a loose-leaf notebook format in order to accommodate additions and revisions easily.
- Many plans have bright-colored covers making them highly visible on a bookshelf should the need arise to refer to them quickly in an emergency. For example, the Baltimore County Public School District uses the same bright orange for its cover that is used for highway warning markers.
- Several plans use a step-fold format with one section of the plan printed on each step. Use of different colored paper for each step helps to distinguish the different sections of the plan and enhances its readability.
- A few school districts publish a generic plan with instructions for individual schools to tailor it to their specific building and to use only those parts of the generic plan that fit their needs.

- A general policy statement from the superintendent or principal; pupil dismissal procedures (including transportation arrangements); and checklists for the principal, teachers, and custodians are common features in the plans.
- Included most commonly in the plans are emergency procedures for such crisis situations as inclement weather, tornadoes, earthquakes, floods, hazardous chemical spills, explosions, nuclear power accidents, plane crashes, assault and battery incidents, power outages, and civil disturbances.
- Less common in the plans — but no less vital — are sections dealing with site emergency management centers, nurse/first-aid centers, hold harmless agreement forms for use when children are assigned to temporary emergency shelters owned by private citizens, a school building floor plan, and a priority checklist to govern actions in an emergency.
- Most plans include very specific procedures for informing the public about emergencies through the mass media. A consistent recommendation is that a single spokesperson be designated for the school or district to collect, coordinate, and release information to the media as it becomes available.
- Most plans include sample letters to parents. The general theme of the letters is that the school affirms its responsibility to attend to the general health and welfare of its students. In case of an emergency, children will remain at school and be cared for by the school staff; or if the emergency limits the use of the school, children will be relocated to an alternative site or dismissed. Also, the letters instruct parents to listen to the radio or to watch television for information and not to call the school (since the telephone would be needed for emergency use) and not to come to school.
- Most school districts file copies of their crisis plans with appropriate community agencies both to keep them informed and to help them devise their own plans to assist the schools should the need arise.
- A few school districts set up crisis management teams to

deal with students, their families, and faculty when a tragedy such as a student suicide or accidental death occurs. Typically, crisis team members include the superintendent, principals, counselors, and certain community members, such as clergy and mental health professionals whose expertise could be useful.

- As a result of the Three Mile Island nuclear reactor incident 10 years ago, Pennsylvania enacted a law that requires all nuclear power plant operators to employ an engineering firm to develop an emergency response plan for every school district and municipality within a 10-mile radius of each plant. The plan must be updated annually, and parents residing within that area must be notified of the plan. While that plan is specific to a nuclear accident, it contains details similar to those found in many of the good general crisis plans that are available.

- Some schools have adopted a "controlled access" policy, because of some tragic incidents that have occurred over the past decade. For example, in Winnetka, Illinois, all school doors, except for the main entrance to each building, are locked during the school day. Visitors are required to enter by the main entrance and sign a visitor's log in the school office. Visitors are issued passes that are to be in their possession while they are on the school premises and returned to the office when the visitor leaves the building. Visitors are escorted to the room or place they wish to visit. Certain individuals are issued permanent passes. Staff are instructed to note if strangers are displaying a visitor's pass and, if not, to escort them to the office. Adults, not students, are on duty in the school office during the lunch period.

- Students bringing weapons to school are an increasing concern. In Duval County Schools in Jacksonville, Florida, school officials developed a plan authorizing security personnel to stake out the entrance of a middle school or high school on a random basis. As students approach the school entrance, they are searched with hand-held metal detectors. This school district reports that these random searches are deterring stu-

dents from bringing weapons to school.
- Essential to any security plan is an immediate communication system. The National School Safety Center recommends equipping schools with two-way walkie-talkie radios for those who supervise children in remote locations, such as the bus boarding areas or the school playground. Also, schools should have two-way call systems between the office and the classrooms.

Five Steps to a Safe and Secure School

Improving school security is a process that begins with an enabling law by the state or a policy adopted by a local board of education. Ultimately, however, the process must focus on the individual school building and the people in it. The following steps are suggested as guidelines for those charged with the responsibility of providing a secure school.

1. *Plan Now.* Have a plan in place before a crisis occurs. The adage, "Don't lock the barn after the horse is stolen," is most applicable to matters of school security.
2. *Do a Security Audit.* Tracking discipline problems, school crimes, neighborhood crimes, weapons in the school, and other hazardous conditions can help a staff to assess the risk factors in a school. The National Alliance for Safe Schools has developed school risk profiles. By applying local audit data to a risk profile, a staff has a basis for its initial efforts to make a school safe and secure.
3. *Involve the Staff.* The entire school staff, including custodians, must recognize that effective education can occur only in a safe environment. Security in a school is everybody's business, not just on fire drill days but every day. Involving staff in the development of a school's security plan and in

assessing its effectiveness generates the awareness and commitment necessary to create a safer school.
4. *Involve Your Community.* Parents have a vital interest in a safe school, even if security procedures sometimes necessitate minor inconveniences for them. Local law enforcement officials have both expertise in security and a strong interest in keeping their jurisdiction safe. Involving representatives of these two groups in the development of your school's security plan will go a long way toward gaining acceptance of and support for the plan.
5. *Keep Up to Date.* Rapid changes in the security needs of a school can make today's plan outmoded tomorrow. It is essential that the plan be reviewed annually and revised, if necessary. Administrators must keep up with the latest effective practices recommended by the security industry. Keeping up to date means continuing training so that faculty and students are always prepared in the event of an emergency.

In the long term, there is no substitute for constant vigilance in matters of school security. Today's circumstances require that school leaders develop a plan to ensure a safe school where children can be both educated and protected. Such a plan must prepare both staff and students to deal with the unexpected.

Resource Addresses

Center for the Prevention of School Violence
20 Enterprise Street, Suite 2
Raleigh, NC 27607-7375
1-800-299-6054
(919) 515-9397
http://www.ncsu.edu

Center for the Study and Prevention of Violence
Institute of Behavioral Science
University of Colorado at Boulder
Campus Box 442
Boulder, CO 80309-0442
(303) 492-8465
http://www.colorado.edu/cspv

National Alliance for Safe Schools
P.O. Box 290
Slanesville, WV 25445
1-888-510-6500
(304) 496-8100
http://www.safeschools.org

National Resource Center for Safe Schools
Northwest Regional Educational Laboratory (NREL)
101 S.W. Main, Suite 500
Portland, OR 97204
1-800-268-2275
http://www.nwrel.org/safe

National School Safety Center
141 Duesenberg Drive, Suite 11
Westlake Village, CA 91362
(805) 373-9977
http://www.nssc1.org

Additional resources can be found on the "Safe Learning Communities: Strategies and Resources" page of the North Central Regional Educational Laboratory website at http://www.ncrel.org/sos/data.htm.

Appendix

The Appendix contains three items: 1) a sample generic policy statement, 2) sample generic regulations to accompany the policy, and 3) a sample form for recording information when a school receives a bomb threat telephone call. These items are offered only as examples and should be adapted to fit local school/community situations.

Sample Generic Policy Statement*

Board of Education _____ School District _____

Emergency at School/s

Purpose: The Board of Education recognizes that its responsibility for the safety of students extends to possible natural and man-made disasters and that such emergencies are best met by planning and preparedness.

Authority: The Board authorizes a system of emergency preparedness which shall ensure that:

1. The health and safety of students and staff are safeguarded.
2. The time necessary for instructional purposes is not unduly diverted.
3. Minimum disruption to the educational program occurs.
4. Students are helped to learn self-reliance and trained to respond sensibly to emergency situations.

*Excerpted from policy suggestions disseminated by Neola, Inc. Used by permission.

All threats to the safety of the school district's facilities shall be identified by appropriate personnel and responded to promptly in accordance with the plan for emergency response.

Responsibility: The superintendent and/or his or her designate shall develop a plan for the handling of emergencies, which includes a plan for the prompt and safe evacuation of the schools, if necessary.

References: List here any statutory requirements of your state.

Sample Generic Regulations*

_____ School District

Emergency Procedures

The primary consideration in any emergency situation must be the safety and welfare of the students and staff. At certain times, therefore, it may be necessary to ask the staff to perform "beyond the call of duty" in order to provide for the safety and welfare of the students. In the event of an actual emergency situation, all school personnel, instructional and noninstructional, are required to remain in the building until they are dismissed by the principal.

A. When an emergency occurs, the principal shall consult with the _____ on whether or not to evacuate the school. An announcement will then be made to inform staff and students of the emergency procedures that will be followed.

B. If a decision is made to send students home or to another location, every attempt will be made to notify parents by telephone and/or by radio or TV.

The professional and support staff members will supervise the orderly evacuation of the building. The teachers will remain in

*Excerpted from regulation suggestions disseminated by Neola, Inc. Used by permission.

(or return to) their regular classrooms and remain with their students.

- ☐ Administration and guidance counselors will report to the main office for assignment.
- ☐ The head custodian will assign the members of the custodial staff to positions that allow "free flow" of traffic on and off school property.
- ☐ Clerical personnel will remain at their regular stations except in cases of reassignment by their immediate supervisor.
- ☐ School lunch personnel will remain in their areas until notified to the contrary by the main office.

Above all, it is imperative that all personnel remain calm and in control throughout the emergency.

Fire Drills

In accordance with state law, fire drills are to be conducted periodically, not less than once a month. Each principal shall prepare and distribute fire drill procedures whereby:

- ☐ All personnel leave the building during a drill.
- ☐ The plan of evacuation provides at least one alternative route in case exits or stairways are blocked.
- ☐ When the fire alarm sounds, teachers caution students to walk silently and briskly from the building to a specified location at least _____ from the building, close all windows and doors of a room and turn out lights before leaving, and conduct roll call as soon as students are out of school and in place to make sure that each student is accounted for.

Unusual Situations

The following situations are quite unlikely to occur; but in the event one of them does, use the following procedural guidelines:

Student or staff member held hostage

- ☐ Isolate the area. DO NOT make an announcement or sound the fire alarm. The nearest administrator will direct teachers and monitors on duty to notify teachers in nearby classes to take their students to a previously determined area.
- ☐ Notify the police immediately, giving as much information as possible (e.g., number of terrorists, number of hostages, types of demands being made, etc.). Police will be in charge once they arrive.
- ☐ Notify the superintendent.
- ☐ Develop a list of casualties, if any.

Intruders in building or on school property

- ☐ Ask them to leave.
- ☐ If they do not leave, remind them of trespassing law.
- ☐ Notify administrator in charge and the central office.
- ☐ Avoid any physical conflict or loud verbal altercations.

The administrator or person in charge may notify police if intruders do not leave and/or summon all "free" teachers.

Demonstrators or pickets around building before school opens

- ☐ Note procedures for intruders.
- ☐ Attempt to enter building peaceably.
- ☐ Notify district office.
- ☐ Hold students on the school buses if demonstrators seem to pose a threat.

Demonstrators or pickets around building at dismissal

- ☐ Follow procedures for intruders.
- ☐ Notify administrator in charge.
- ☐ If demonstrators seem to pose threat, hold students in class until further notice.

Student demonstrators

- ☐ Follow above guidelines.
- ☐ Identify the leader or leaders of the group.

- ☐ Notify the group that the administration will confer with the leader(s) but not until all other students return to class.
- ☐ Meet with leaders if students disperse and return to class.
- ☐ If students refuse to disperse, remind them of truancy regulations and consequences of disruption.
- ☐ With other staff, try to identify as many participants as possible; notify parents of participants, asking them to come to school.
- ☐ Try to keep nonparticipating students away from the demonstration area.
- ☐ Follow through on disciplinary action for students refusing to cooperate.

Buses not operating — students in school

- ☐ Keep all students in class.
- ☐ Summon all "free" professional staff members for other assigned duty.
- ☐ Dismiss students with their own cars (or bicycles), but follow board policy regarding student passengers.
- ☐ Dismiss students who live within walking distances after the cars have left.
- ☐ Keep remaining students in school until parents are notified of the problem and arrange transportation for their children.
- ☐ Alert school lunch staff if it appears that large numbers of students will have to remain at school for an extended period of time.

Major disruptions at an assembly

- ☐ Bring house lights to full on.
- ☐ If those causing the disruption are too numerous to be handled by professional staff members on duty, dismiss the assembly and instruct all present to report to their next period classes.
- ☐ If few in number, remove the disruptive students from the assembly and take appropriate disciplinary action.

- [ ] If those causing the disruption are outsiders, follow the procedure for intruders.

Emergency Bus Evacuation

Each bus driver in conjunction with the _____ shall conduct at least two (2) bus drills per year with each bus load of students. In the event of a real emergency, the following procedures shall be used. Drills are to be patterned on this procedure:

- [ ] Stop bus, if not already stopped, off the road and out of the mainstream of traffic, if possible.
- [ ] Put on emergency flasher lights and immediately issue orders as to which exit is to be used.
- [ ] Maintain calm and direct students to "walk, not run" and to watch their heads when disembarking.
- [ ] Send someone for help immediately, after making sure that all students are evacuated.
- [ ] Form students into a group at least 100 feet from the bus and as far away from traffic as feasible.

Bomb Threats

The majority of bomb threats are hoaxes and result in nothing more than a disruption of the school routine. However, the chance remains that the threat may be authentic; and appropriate action should be taken in each case. To help prevent bomb threats, teachers should briefly check their classrooms on arrival each morning and upon returning to their classrooms during the day and report to the principal or custodian any unusual circumstances or articles left in the room. Custodians should lock doors when they leave a room after cleaning, if during non-school hours.

- [ ] If a bomb threat is phoned into the school, attempt to delay the caller and to obtain as much information as possible. Do not hang up the telephone. In this manner the call can be traced. Someone will then phone the police on another line.

- [ ] The public address system will be used to inform all teachers and students that:
    1. All students will go to their lockers, unlock them and take a coat if the weather requires, and return to class—leaving locker doors open.
    2. The standard fire drill procedure will be used to evacuate the building, if necessary.
- [ ] Along with the police department, the _____ will search the building.
- [ ] The importance of good student control is critical while the search is being conducted. Students should remain with their teachers, and those teachers who are not directly responsible for a class at that time should be in the area to help maintain the desired degree of control.
- [ ] If an explosion should occur, students should be moved in the following manner:
    1. Elementary students will move to _____
    2. Secondary students will move to _____

Bomb Threat Check List*

Call received by _____ Time_____
Date_____

A. Ask these questions:
   Where is the bomb? _____
   When will it go off? _____
   Why are you doing this?_____
B. Evaluate the voice of the caller and check the appropriate spaces:
   Male ____ Female ____ Child ____ (approximate age)
   Appears intoxicated ____ Speech impediment ____
   Other special characteristics (describe) _____
   _____

C. Listen for background noise and check the appropriate spaces:
   Music ____ Babies or Children ____ Machine Noise ____
   Conversation ____ Cars/Trucks ____ Airplanes ____
   Typing ____ Other ____
D. Remarks: (Give exact wording of threat message)
   _____
   _____